YOUR KNOWLEDGE HAS VALUE

- We will publish your bachelor's and master's thesis, essays and papers

- Your own eBook and book - sold worldwide in all relevant shops

- Earn money with each sale

Upload your text at www.GRIN.com and publish for free

Martin Wendt

Derivatives on Agricultural Commodities. Harm or Charm for the World's Society

GRIN Verlag

Bibliografische Information der Deutschen Nationalbibliothek:

Die Deutsche Bibliothek verzeichnet diese Publikation in der Deutschen National-
bibliografie; detaillierte bibliografische Daten sind im Internet über http://dnb.d-
nb.de/ abrufbar.

Imprint:

Copyright © 2014 GRIN Verlag GmbH
Druck und Bindung: Books on Demand GmbH, Norderstedt Germany
ISBN: 978-3-656-66852-7

This book at GRIN:

http://www.grin.com/en/e-book/274334/derivatives-on-agricultural-commodities-
harm-or-charm-for-the-world-s

Derivatives on Agricultural Commodities

–

Harm or Charm for the World's Society?

Home Assignment in the Elective Finance

submitted by

Martin Wendt

to the Department of Finance & Accounting of HSBA Hamburg School of

Business Administration

Date of Submission: January 6th, 2014

<u>Abstract</u>

The following home assignment deals with the question whether derivatives harm society as a whole. The central issue is a possible impact of derivatives prices on present prices for agricultural commodities. In order to provide the necessary background knowledge, the terms derivatives, varieties, valuation, market participants and principles of operation are explained. After recognising theoretical knowledge and the results of different studies, it turns out that there is no empirical evidence for plausible harming impacts of over speculating index investors. Further studies have to be conducted in future in order to provide a reliable proof for policy makers.

Table of Contents

<u>Tables</u>

Abbreviations

FAO	Food and Agriculture Organization of the United Nations
Forwards	Forward Contract
Futures	Futures Contract
IWIM	Institute for World Economics and International Management
OTC-markets	Over-the-counter Markets

1. Introduction

The world's society has been faced with price increases in different spheres of life in the last decade. Among other impacts, people have to pay higher rents in metropolitan areas and have to adjust their monthly budget to higher energy costs. Especially the public debate on increased food prices is very prevalent since people in third world countries are faced with undernutrition and death from starvation. According to the *Food and Agriculture Organization of the United Nations* (FAO) about 8,5 % of the world's population suffered from chromatic hunger in the last years (FAO 2013a). A recent meta-study of the *Institute for World Economics and International Management* (IWIM) led by Hans-Heinrich Bass (Bass 2013, p. 16) heated up the debate in the last two months of 2013 because it claims that derivatives have contributed to the price development with a high probability. In Germany the *Deutsche Bank* and the insurance company *Allianz* are involved in trading with the particular derivatives. Therefore they are heavily criticised by organisations like *Foodwatch* and *Oxfam*. Furthermore even the German president Joachim Gauck, as head of state, spoke out in favour for the critics. He aims to prohibit speculation with agricultural commodities (Oxfam 2013). On the other hand scholars like Don M. Chance and Robert Brooks (2010, p.14) state about derivatives that: "Society benefits because the prices of the underlying goods more accurately reflect the good's true economic values." So who is right and who is wrong and on which evidence do the schools of thought base their implications? In order to give theoretical background to this complex public debate this assignment gives answers to the following questions:

What are derivatives with respect to agricultural markets? What are market participant's intentions? How derivatives work and what is their value? Central question: Are derivatives on agricultural markets a harm or a charm for society? After giving a technical definition about derivatives, an overview about relevant types of derivatives and their connection to agricultural commodities will be given. Afterwards their principle of operation is illustrated with the help of a practical example. For the purpose of being able to discuss the potential impacts of derivatives on food prices, the theoretical foundations about valuation, types of traders and the technical relationship to prices are explained furthermore. The conclusion of chapter 6 delivers an interim result for the answer to the central question.

2. Derivatives Defined

When talking about the definition of derivatives the best way to get the actual meaning is to refer to the root word "derive". Derivatives as financial instruments are assets whose valuation derives from the performance of an underlying asset (Chance and Brooks 2010, p. 1). Hence, the return on a derivative instrument completely depends on the value of a second variable (Hull 2012, p. 1). Concerning the definition, the scholarly literature agrees on this rather technical explanation. The main transaction (exchange of an underlying asset) has to be seen separate from the financial instrument. Derivatives were initially shaped as professional traded assets around agricultural commodity markets in Chicago in the 19th century (Andersen 2006, p. 127). The underlying assets are prevalently tradable assets, these are in agricultural markets for instance corn, soybeans, wheat, sugar or cocoa.

Whether operating in agricultural markets, stock markets or other commodities markets, etc., the buyers of derivatives can have diverse intentions by holding such a financial asset. Before going into detail with the different types of buyers in chapter 5, generally the motivation to reduce (also referred to hedging) a financial risk can be assumed (Hull 2012 p. 1-2). The risk, that is to be secured, arises out of the price volatility of any tradable asset (Chance and Brooks 2010, p. 5). A quite practically relevant example is the variation of exchange rates for foreign currency. Companies doing international business fear the risk of fluctuating currency rates in the period between input expenditures and final payment. Characterized by a value-dependence on a second variable (here the exchange rate) a derivative instrument provides protection and has therefore developed to an often-used tool among companies (Chance and Brooks 2010, p. 12). In other words, possible risks can be transferred from one party to another (Bösch 2012). That's why derivatives have undergone phenomenal growth since the 1970s (Jarrow and Chatterjea 2013, p. 4).

3. Types of Derivatives

After clarifying the basic characteristics of derivatives in general, a deeper look into different types is needed to assess their usage on agricultural markets. A typical risk transfer on these markets could be undertaken in order to secure the risk of falling or rising prices for commodities.

For instance, a farmer wants to fix a certain price for corn that will be delivered to the buyer in six months after the next harvest. In this case suitable tools are called *forward contract (also called forwards)* or *futures contract (futures)*. This chapter therefore aims to give an overview about these two relevant instruments.

3.1 Forwards

As shown in the example, a forwards is an agreement between two parties to buy or sell a certain asset at an agreed time in the future to a certain price (Hull 2012, p. 3). Unlike options (right to sell or buy an asset) the agreement reflects a binding commitment or in other words an obligation for both the delivering and the paying party. The arranged price for the transaction is called the *forward price* or *delivery price.* Just the opposite, the current price of the equivalent asset is called the *spot price.* The difference of both prices of the underlying asset between agreement and delivery is the value that forwards holders derive their profit or loss from (Jarrow and Chatterjea 2013, p. 212-224).

Different to stocks or options, forwards are not traded at official exchanges. Typically there are traded via a direct communication line among financial institutions such as banks in a less formal way, called over-the-counter markets or OTC-markets (Chance and Brooks 2010, p. 253).

3.2 Futures

More common for underlying agricultural commodities are futures being traded in a more organised manner on futures exchanges. Just the same as a *forwards, a futures* corresponds in theory to an obligation for the seller and the buyer for an exchange of assets on a delivery day to a delivery price in an agreed quality. The quality is fixed by delivery terms and contract specifications. Having developed out of forwards, futures are more liquid contracts since holders can easily trade them and thus can be straightforwardly relieved from their obligation to sell or buy the underlying asset. The more technical term for selling that obligation is closing an open position or offsetting. What is more is that due to every day spot price fluctuations of the commodities, the agreed forward price gains or loses attractiveness for both the derivative holders. The daily wins and loses of this zero-sum game (one's gains on the expense of the other party) are subject to a regular cash settlement (Chance and Brooks 2010, p. 3).

In essence both the forwards and the futures are comparable to a bet between two market participants. The obliged future buyer of the asset expects increasing prices and wants to fix the price (delivery price). The seller, expecting falling prices, wants to gain profit by supplying the asset on the delivery date for a lower price than the delivery price. As both have an interest on such a bet, they meet each other on derivative markets in order to agree on a buying/selling obligation.

4. Conceptual Insights and Valuation

As the mechanisms of futures and forwards might sound abstract at first glance, this chapter provides a more practical approach to their usage, underlying mechanisms and valuation based on the *carry arbitrage model*. Prior to the following example it can be concluded that there are significant similarities of both previous described types. Therefore a deeper view is taken into the actively traded and in agricultural sectors commonly used futures.

4.1 Main Mechanisms of a Futures

A farmer A owns several cornfields and actively trades with his harvested corn. The spot price of a so-called bushel is USD 4,282 on the 19th of December 2013 (Nasdaq 2013).

Figure 1: Spot Prices on Corn in Cents per Bushel[1]

Figure 1 shows the development of the corn prices on the spot market during the last six months. As most individuals are risk averse, A also wants to transfer the risk that he is facing due to falling prices (Chance and Brooks 2010, p. 7).

Hence, A is willing to step into an obligation to sell his corn in May 2014 for a fixed price, in other words a six months futures. Technically the obligation to sell or deliver is called a short position (Hull 2012, p. 102). The delivery price is for the most part determined by the costs of carry of the physical asset until delivery

---

[1] Nasdaq 2013.

(explained in detail in chapter 6). Both A and B virtually meet each other at a futures exchange and arrange USD 4,382 as the delivery price per traded bushel. Besides the price, a quantity of 100.000 bushels and a certain quality standard are arranged. For further assessment of the futures, a spot price on the delivery date of USD 4,48 is assumed (see table 1). The moment where they fix the contract is referred to as t_o.

Spot price (S)	Futures price or delivery price (K)	Spot price on delivery date (S_t)
USD 4,282	USD 4,382	USD 4,48

Table 1: Overview on Prices[2]

As stated in chapter 3.2 a futures can be described as a zero-sum game. In this example B gains the difference between S_t and K ($S_t - K$) on the expense of A. In other words B is on the long position and has the obligation to buy and due to an increased price S_t has the advantage to buy for the price K below market price, whereas short position seller A could have sold for S_t. He therefore loses $S_t - K$ in the amount of USD 9.800 as shown in table 2.(Chance and Brooks 2010, p. 270).

Contracting Party	Farmer A	Trader B
Intention	Sell on May 15th 2014 own corn due to limited warehouse capacity	Need to buy corn to deliver it to his customers on May 17th 2014
Expectation	Falling prices	Increasing prices
Position	Short	Long
Value of the futures on delivery date	100.000 bushels x (USD 4,382-USD 4,48) = - USD 9.800	100.000 bushels x (USD 4,48 – USD 4,382) = + USD 9.800

Table 2: The Corn Futures between A and B and its Result[3]

In this example the prices have increased in favour for the trader B. Since the outcome of such a futures is uncertain, infinitely different win-loss combinations can occur. The following figure 2 shows how the payoffs from futures/forwards can be distributed.

---

[2] Own example.
[3] Own example.

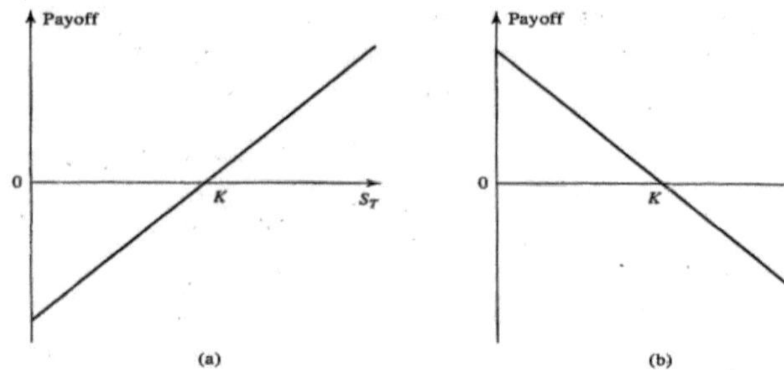

Figure 2: Payoffs from Futures/Forwards[4]

(a) long position, (b) short position, delivery price = K, price of asset at contract maturity = S_t

Being on the long position Trader B' s possible outcomes are represented by diagram *(a)*. A positive payoff for B appears as long as the spot price at the maturity S_t is higher than the fixed delivery price *K*. The contrary holds for A represented in diagram (b). If *K* is equal to S_t no party can reap a profit from these derivative instruments.

4.2 Valuation Concept

The question at this point is how the gain or the loss of the contracting parties materialises or to put it different, whether and how positive or negative cash flows occur during duration and on maturity.

Generally just a few derivative traders hold their contracts until the maturity date (Chance and Brooks 2010, p. 12). The reason why about 99 % of all futures are not delivered, comes from the two layers of a derivative (Chance and Brooks 2010, p. 274). The general definition in chapter 2 has claimed a connection between the value of the derivative and an independent variable. Furthermore the derivative contract technically leads to the fulfilment of an underlying transaction on maturity (here the delivery of corn). To sum up the pure derivative and the underlying transaction have to be seen separately. Thus they are divisible and carry their own intrinsic value. Switching the role of B from a trader, that has an interest on the real delivery of corn (on maturity), to a profit seeking investor, he could have offset his obligation before maturity.

In order to discover the fundamental value drivers of futures, transaction costs coming from commissions, delivery and market illiquidity have to be neglected

---

[4] Hull 2012, p. 6.

(Chance and Brooks 2010, p. 281). Moreover a narrower distinction between value and price of a contract needs to be kept in mind. Actually the concept of strong market efficiency states that due to perfect information, available to all market participants, the price of an asset equals its value. For futures this concept must be seen with some adjustments since from a long position perspective the purchase of the futures itself is costless. Notwithstanding that a future cash flow from reselling is expected (Chance and Brooks 2010, p. 288). When buying the futures from A in t_o, no initial price will be charged from B. Aiming to reduce the risk of non fulfilment due to illiquidity on maturity (credit risk), futures exchanges require small initial margins (deposits of 5-7 % of the contract value) from both parties. As mentioned in chapter 3.2 futures are subject to a regular cash settlement to distribute daily profits and losses. They result from everyday spot price fluctuations causing as well futures prices to fluctuate. Their relationship will be explained in chapter 4.3. For instance, one day after A and B had arranged USD 4,382 as K, the spot price went up causing the futures price to rise as well. The amount multiplied by the fixed quantity of bushels will be transferred from margin account A to B. The settlement runs in that direction since long position B is in favour of increased spot and futures prices. This daily procedure is also called marking to market. Besides the fact that there is no initial price in t_o, it can be deduced that the value of the futures in t_o equals zero.

Another implication of the *carry arbitrage model* is that the futures price at expiration must equal the spot price. Otherwise selling spot contracts and buying futures or vice versa would provide risk-free profits in other words arbitrage profits. This is based on the assumption that the same good at the same time is traded for one price all over the world neglecting transaction costs, inflation and delivery costs (law of one price). (Chance and Brooks 2010, p. 288)

Being marked to market at the end of every trading day, the opening price of a futures a the current day would equal the settlement price of the prior day. Price changes during the trading day prior to daily settlement would cause a change in contract value. As soon as the futures is marked-to-market its price reflects an *adjusted spot price* of that trading day, gains or losses are distributed among contractors and the value reverts back to zero (Chance and Brooks 2010, p. 292). This *adjusted spot price* charges the so-called cost of carry, which will be explained in the next section. In general the positive value of a futures at any point of time is equivalent to the profit that could be realised by selling it at this point of time (Bösch 2012, p. 150). Referring to the given example, the value for B at any given point of time equals the balance of all price movements until that point of time (multiplied with the agreed quantity). This would be the same if both parties would have agreed on a forwards, because the price of both derivative instruments is equal if no credit risk and thus no marking to market is assumed. Otherwise the higher default risk of forwards and the possibility of realised daily profits for futures holders and their costs of capital would make the difference in pricing (Chance and Brooks 2010, p. 294).

4.3 Relationship between Spot Markets and Futures Markets

In fact, already spot prices incorporate market participants' expectations of further price developments. If the farmer A in t_0 would have been able to foresee a hard winter resulting in a future undersupply of corn, then all other market participants would have been able to. This argument is driven by the assumption of efficient markets (no arbitrage opportunities) and leads to higher spot prices recognising future undersupply (Bösch 2012, p. 151-152). Theoretically and under certainty only spot prices can influence futures prices but not vice versa. The reason for that is a strong one-way relationship between both prices defined by the costs of carry.

Basically there always has to be a difference between spot and futures prices prior to the expiration since the date of fulfilment simply differs. A purchase on the spot market is always associated to costs of carry compared to a later procurement of the agricultural product (Chance and Brooks 2010, p 291). Unlike stocks, commodities do not provide positive cash flows during storage and their costs of carry are composed of financing and storage costs. Hence, the price of a futures -ceteris paribus- has to be higher than the spot price if the costs of carry are positive (Bösch 2012, p. 151). Under certainty the price for a futures is calculated the following way[5]:

$$(1) \quad f = S_0(1+r)^T$$

S_o= Spot price
r = Risk-free interest rate
T= Time to maturity in days

To sum up, it is the spot price compounded to the date of maturity at a risk-free interest rate.

---

[5] Chance and Brooks 2010, p. 291.

The risk-free interest rate *1+r* as a positive value is derived from the positive sum of all carrying costs (financing costs and storage costs). Exponentiated by the outstanding days and multiplied by the spot rate, it leads to an amount of the futures price which is higher than the spot price. That reveals that no additional value can be generated by simply creating a futures of a generic asset tradable on the spot market. The difference is just driven by the costs of carry and the period of time until fulfilment. The prior mentioned hypothesis that the initial value of a futures is zero and that the futures price and spot price converge on maturity therefore holds as shown in Figure 3.

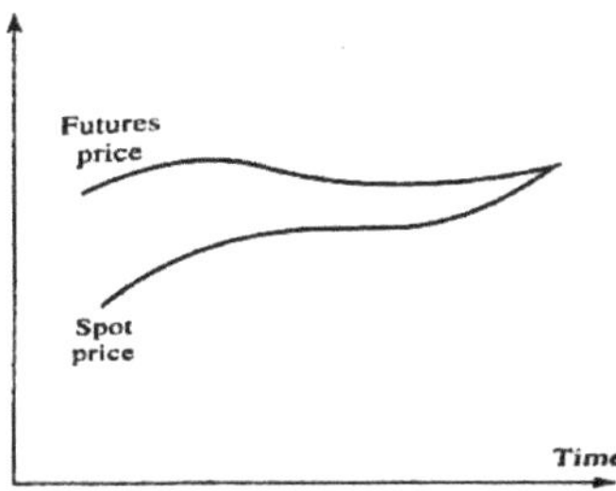

Figure 3: Convergence of Futures Price and Spot Price[6]

5. Markets, Participants and their Intentions

After having clarified the characteristics of the relevant derivative instruments, their principle of operation as well as their valuation, a deeper view on markets and participants will be given. For the purpose of assessing whether social value is created or carried away by derivative instruments in agricultural markets, a special focus is made on intentions and strategies of derivative traders. Furthermore the question, whether they can all be called society harming gamblers or not, will be answered in this chapter.

Basically four different groups of traders act on derivative markets. The first group was already mentioned as the role of Farmer A and is called *hedgers*. His intention was to reduce his risk that could have resulted by decreasing prices (Hull 2012, p. 12). With the help of derivatives, hedgers seek to transfer the risk to another party. The group, which is willing to take risk and seeks in exchange

---

[6] Hull 2012, p. 28.

for compensation, is called *speculators*. Important to note is that speculators primarily try to use their experience in price developments in order to place a bet, which is a profit opportunity to them. To sum up they hold capital and expose it consciously to a risk that naturally exists like a price risk (Bösch 2012, p. 7-9). *Market Makers* typically do not want to take risk and just bring together interested sellers and buyers. Whenever they buy or sell a derivative they enter an open position that is intended to be closed as soon as possible. Therefore profits are mostly made up of a trading margin but are also exposed to a risk when holding the derivative (Bösch 2012, p 9). Market participants, who strictly want to avoid risk and aim to realize riskless profits without holding capital, are called arbitrageurs. They simply use price differences of assets on different markets that exist only for a short timespan. Thereby they play an important role for the overall market since they establish market efficiency when they force different prices to converge with their trading (Andersen 2006, p. 136). Taking advantage of sometimes hidden short-time market inefficiencies, arbitrageurs as professional participants are specialised in identifying inefficiencies with special tools (Bösch 2012, p. 9).

The previous mentioned markets participants can either act on OTC-markets or on exchanges like the Chicago Mercantile Exchange emerged in 1919. The essential difference is that the less regulated and less standardized forwards are fixed bilaterally between financial institutions. Due to possible individual arrangements they are less attractive to third parties and hence not likely to be traded. Official exchanges can be accessed by a wide range of market participants as there are regulated and offer trade of standardised derivative contracts like futures. At first glance, none of the trader groups could be called pure gamblers since they all serve the purpose of establishing and balancing a market for risk transfers demanded by risk avoiding members of society. Following the basic definitions even the speculators can simply be renamed as risk takers that are essential for the existence of derivative markets. Whether the theoretical approach of the previous chapters matches practical insights and whether reliable evidence for undesirable development exists, will be covered in the next chapter.

6. Derivatives on Agricultural Markets and their Impacts

After having built a theoretical framework for discussing the main reproach that trading with derivatives partially contributes to increasing food prices, an analysis will be given in this chapter (Bass 2013, p. 8). As a first step the peculiarities of food markets with regard to price volatility will be shown in order to distinguish between normal and abnormal fluctuation.

6.1 Market Characteristics

Typically underlying assets of agricultural markets are corn, wheat, soybeans, rice, cocoa, coffee, cotton or products related to livestock. A lot of these products are input resources for groceries. So basically the two forces supply and demand lead to a price, which is perfectly balanced in the equilibrium of an efficient market (Mankiw and Taylor 2008, p. 76). If there is overdemand on a product, for instance through a rise in population or income, the price will go up. The same will happen if there is undersupply leading to a shortage. Moreover the prices are influenced by a *stocks-to-use ratio,* that reflects how much of resources are in stock or in other words extracted from the active market. Additionally, prices are subject to seasonal fluctuation through changes of weather or storage costs. Effects of inflation as a general rise of price levels are neglected (Chance and Brooks 2010, p. 275).

6.2 Analysis of the Link to Increased Food Prices

As stated in the introduction a current meta-study of the IWIM comes to the result that there is a high probability that price developments on futures/forwards markets influence spot markets (Bass 2013, p. 16). Both markets are said to be correlated via arbitrage opportunities in both directions according him. For instance, if a futures price is too high, an arbitrageur (as a mechanism of efficient markets) will have an incentive to buy on the spot market and enters a short position to realise profits on maturity from the price difference. It is at least plausible that shifting the demand from the future to the present time, because of price incentives, has an influence on the spot market. Nevertheless it contradicts to the one-way relationship stated in the carry arbitrage model. As investment opportunities in real estates were not attractive to investors anymore and expansionary monetary policy of both the Federal Reserve Bank and the European Central Bank started in 2007 (IMF 2013, p. 4), additional liquidity has begun to spread across the globe. Seeking for return on investment, speculators entered futures markets for agricultural commodities causing trading volume on these derivative instruments to rise sharply in 2008 (Masters and White 2008, p.1). At the same a strong increase in prices of agricultural products can be observed. This has been measured by the FAO with the help of the price variation of a consumption basket leading to the so-called FAO Food Price Index (FAO 2013b, p. 131-133). Michael W. Masters and Adam K. White (Masters and White 2008, p. 13) claim that especially index investors put buying pressure on food markets with their long positions resulting in higher spot prices. Index investors are a subtype of the before mentioned speculators. Those institutional investors want to invest into a package of different commodities hoping to reap benefits from increasing prices. Therefore large pensions funds invest in commodity index portfolios that are actively managed by bank traders. As these traders do not want to speculate in a specific commodity, they just follow the overall performance of an index which is composed of different shares of commodities. As they differ in proportion, price movements of certain commodities can have a dissimilar influence of the overall development. For instance, if there is an upwards trend of crude oil prices, traders will have to enter additional long positions in other commodities to keep on track with the index. Being risk averse those investors rely on long-term trends and are called passive investors. They typically have the opinion that over-average profits occur coincidentally (random-walk theory) and hence cannot be realised sustainably due to perfect information on efficient markets (Hillier et al 2013, p. 351-377). This is why they follow the average trend of a certain portfolio. It can be concluded that following upward trends, index investors typically hold long positions and their decision to enter a position does not come from particular knowledge about the development of supply and demand. Technically speaking, index investors that are being faced with criticism from Masters and White (2008) and Bass (2013) at least serve one important purpose of their role as speculators. They supply willingness to take risks from food commodity sellers. Bass (2013) states that the provided insurance is offset by index investors through an excessive flow of capital that distorts a realistic equilibrium price in futures and via arbitrage also to spot markets. The second important task of speculators is the price discovery. This simply means that farmers can use spot prices and futures prices for their negotiations with customers. Furthermore it offers them the possibility to assess whether it is worthwhile for them to sell or stock their good recognising individual costs of carry. Hence, business decisions are taken by recognising futures prices. According to Masters and White (2008,

p. 11) this price discovery function is disturbed by the artificial financial demand caused by cash inflows from institutional investors.

Other than the previous fundamental criticism on institutional index investors, there are other schools of thought, which do not see a correlation between increased trading volume and increased prices. Following Ingo Pies (Pies 2013, p.1) the price development was driven by real economic factors such as increased world population and per capita income, competition on usage of agricultural products for energy production and bad harvests. Furthermore low stock levels as well as restrictions on imports and exports around the globe contributed to higher prices on food. Another study conducted by Nicole M. Aulerich, Scott H. Irwin and Philipp Garcia (2013, p. 37) comes also to the result that index investment was not the reason for the price development in agricultural markets. From their point of view policy makers should strengthen the supply side of the agricultural sector, reduce trade barriers and remove misallocations caused by incentives for the production of bio energy.

As a result of the intensive scientific discourse on this subject, a lot of empirical studies try to answer the question whether correlations can be found. With the help of the granger-causality method, scholars compare two variables X and Y reflecting values of different time series. For example they compare weekly returns on long positions and the amount of open positions of index traders. The results primary reject any possible correlation but unfortunately show lack a of underlying valid input data (Bass 2013, p. 32- 35). As a key finding it can be concluded that the criticism on food speculation is backed up by a plausible model of thinking, but lacks a broad verification with the help of reliable empirical methods. What is more is that price developments are subject to a complex set of interrelating factors. The conducted empirical studies have concentrated on a few influencing factors without reliably isolating them from real economic factors (e.g. stock-to-use ratio). Therefore the question is still not finally answered with regard to full empirical validation and a scientific consensus. Nevertheless in the recent global economic prospects report of the World Bank (2013, p. 110) at least the strong price volatility of the last decade is referred to the financialisation of commodity markets. The global investment fund activity from 2006 to 2012 has risen from USD 140 billion to USD 330 billion (World Bank 2013, p. 109). As shown in the following figure a sharp increase in the agriculture price index can be observed as well.

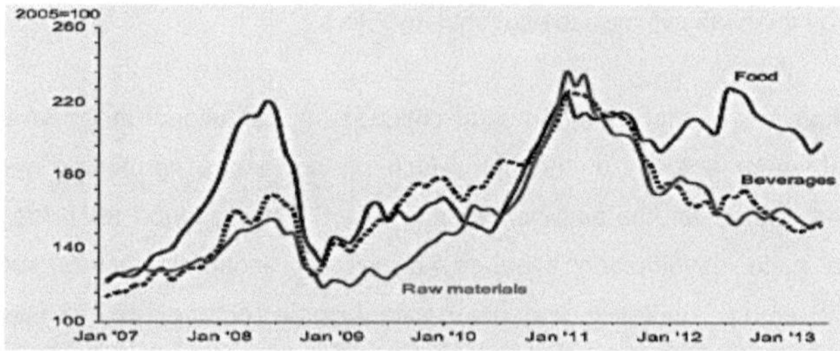

Figure 4: Development of the Agriculture Price Index[7]

6.3 Conclusion

Agricultural commodities are generic goods that can either be stocked or used for immediate consumption. Typically the forces supply and demand lead to an equilibrium market price. Market participants can trade on the spot market or commit themselves to sell or buy in the future. With regard to futures and forwards, demand has to be distinguished in natural (physical) demand and financial (artificial) demand. Spot and futures markets are connected via arbitrage opportunities. In fact, Pies (2013) states that low stock levels contributed to the high fluctuation on spot markets. According to a rational behaviour of traders, stock levels will go up if futures/forwards are overpriced. So what did actually happen with the measurable stock levels? The study of Pies (2013) and Aulerich, Irwin and Garcia (2013) claim that no increase of stocks could be recognised. The apparent contradiction is commented by Bass (2013, p. 16) in return. The input data used to make statements about stock levels was not reliable according to his working paper. Three quarter of the leading corn trading companies would never reveal their stock levels for strategic reasons. Furthermore it has been found out that the previous quoted scholar Irwin is closely related to the American financial industry. This final example of the prevailing academic discord gives the impression that this professional conflict is yet not solved by profound knowledge. The final answer whether a harmful relation between futures/forwards exists, has not been answered empirically due to a lack of an appropriate measuring tool that is accepted by the majority of the scholars. Nevertheless both schools of though agree on the fact that futures and forwards can be useful for price discovery and risk transfer. Creating value for parties that want to transfer their risk, they can contribute to an enhanced social value. Disagreements exists regarding the group of index investors that simply participate in price increases and lead with their long positions to a self fulfilling prophecy.

---

[7] World Bank 2013, p. 105.

7. Result and Critical Appraisal

Coming back to the complex debate on agricultural speculation, derivatives are financial instruments that help spot market participants such as farmers to hedge risk. Futures, as the predominantly traded type, are standardized contracts that enable speculators to take a farmer's risk. The two other markets participants are arbitrageurs and market markers. Both are important to keep the market balanced. Regarding the valuation, it can be concluded that forwards and futures can be valued equally under the before mentioned assumptions. A basic foundation is that spot and futures prices differ from each other in the amount of costs of carry for the particular physical asset holder. As soon as the contract expires, spot and futures prices converge in efficient markets. Otherwise riskless profits could be realized. That's why Bass (2013) claims that both prices are connected via arbitrage opportunities. At first glance this answers the central question and relates the increasing speculative demand of index investors to higher spot prices. Non-physical demand driven by yield prospects contradicts the quote of Chance and Brooks (2010, p. 14) because it apparently does not lead to fair values of physical assets. Nevertheless the counterside argues that there is no reliable evidence and emphasises the results of several studies, which in return is criticised by Bass (2013) with regard to the applied empirical methods and input data. To put it briefly, there is no scientific consensus but a nonconstructive discussion about formalities and empirical measures among the scholars. The impact on prices of physical goods and its bad implications for the social value (decreasing consumer surplus) is quite plausible when comparing prices and investment volumes. It lacks an empirical verification that possibly is not desired by the winners of these zero-sum games. The applied methods have failed to isolate the depended and non-depended variable from real economic impacts. Moreover the analysed data does not reliably consider the investment volume in forwards, since these unstandardized contracts are bilateral agreements that are not recorded accurately. However, trading with derivatives cannot be rejected wholesale due to the mentioned advantages for farmers. They are neither a pure harm nor a pure charm. Based on little capital commitment and high leverage, derivatives are attractive for index investors that have no insights in the physical market. Regulating initial margins could serve the purpose to make investments less attractive to them because of higher capital commitment. In spite of that, further studies should be conducted to work out the yet plausible relationship with empirical verification.

References:

Andersen, Torben J.: Global Derivatives – A strategic Risk Management Perspective. 1st edition. Pearson, Harlow 2006.

Aulerich, Nicole M., Scott H. Irwin, Philip Garcia 2013. "Bubbles, Food Prices, and Speculation: Evidence from CFTC's Daily Large Trader Data Files, National Bureau of Economic Research, working paper 19065, Cambridge/Massachusetts.

Bass, Hans-Heinrich. 2013. "Finanzspekulation und Nahrungsmittelpreise – Anmerkungen zum Stand der Forschung". Materialien des Wissenschaftsschwerpunktes Globalisierung der Weltwirtschaft, Institute for World Economics and International Management, Working Paper Nr. 42, Bremen.

Bösch, Martin: Derivate – Verstehen, anwenden und bewerten, 2nd edition. Vahlen, München 2012.

Chance, Don M. and Robert Brooks: An Introduction to Derivatives and Risk Management. 8th edition. South-Western Cengage Learning, Mason 2010.

FAO 2013a. "The State of Food Insecurity in the World", Rome.

FAO 2013b. "Food Outlook". November 2013, Rome.

Hillier, David, Stephen A. Ross, Randolph W. Westerfield, Jeffrey Jaffe, and Bradford J. Jordan: Corporate Finance. Second European edition. London et al. 2013.

Hull, John C.: Futures, Options, and other Derivatives. Global Edition. 8th edition. Pearson, Boston et al. 2012.

IMF-International Monetary Fund. 2013. "World Economic Outlook". October 2013, Washington D.C.

Jarrow, Robert A. and Arkadev Chatterjea: An Introduction to Derivative Securities, Financial Markets, and Risk Management. 1st edition. W. W. Norton & Company, New York 2013.

Mankiw, Nicholas G. and Mark P. Taylor: Economics. 4th edition. South-Western Cengage Learning, London 2008.

Masters, Michael W. and Adam K. White. "The Accidental Hunt Brothers: How Institutional Investors are Driving up Food and Energy Prices." 2008, http://www.loe.org/images/content/080919/Act1.pdf (accessed December 29th , 2013).

NASDAQ 2013, Corn Daily Price,http://www.nasdaq.com/markets/corn.aspx (accessed December 20th , 2013).

Oxfam 2013. "Bundespräsident Gauck spricht sich gegen die Spekulation mit Nahrungsmitteln aus." http://www.oxfam.de/news/121218-bundespraesident-gauck-spricht-sich-gegen-spekulation-nahrungsmitteln (accessed January 2nd, 2013).

Pies, Ingo. 2013. "Agrarspekulation? – Der eigentliche Skandal liegt woanders!". Faculty for business economics Martin-Luther-university of Halle-Wittenberg, working paper nr. 2013-4, Halle.

World Bank 2013. "Global Economic Prospects". Volume 7 June 2013, Washington D.C.